LAKELANDERS

LAKELANDERS

Stories and poems about
living in a Lake District valley

Andrea Meanwell

HAYLOFT

First published by Hayloft Publishing Ltd., 2018

A CIP catalogue record for this book is available from the British Library

ISBN 978-1-910237-46-5

Designed, printed and bound in the EU

Hayloft policy is to use papers that are natural, renewable and recyclable products and made from wood grown in sustainable forests. The logging and manufacturing processes are expected to conform to the environmental regulations of the country of origin.

This book is printed with the offset of carbon emissions by additional carbon offset projects. Supported offset project: Forest Protection, Pará, Brazil

Hayloft Publishing Ltd,
a company registered in England number 4802586
2 Staveley Mill Yard, Staveley, Kendal, LA8 9LR (registered office)
L'Ancien Presbytère, 21460 Corsaint, France (editorial office)

Email: books@hayloft.eu
Tel: 07971 352473

www.hayloft.eu

For the residents of the Rusland and Crake Valleys
Past, present and future.

LAKELANDERS

Contents

Introduction	8
Prologue	9
Certainty	11
Cultural Heritage	13
The Children of the Lake	15
Henry's Owl	17
Joe's Cows	20
Winter in the valley	25
Valley Legacy	33
Lost in Lichen	35
Valley of the Owls	36
The Bridge Over the Stream	38
Celebrity lengthsman	39
In Scruff	40
Hulleter Sheepwash	42
Bethecar Moor	47
Rusland	112
Acknowledgements	114

Introduction

I was lucky enough to live and farm in the Rusland and Crake Valleys from 2009-2017. While there I wrote two memoirs about my time farming, *A Native Breed* (2016) and *In My Boots* (2017). Buying a new farm and moving to it took a long time to accomplish, and instead of writing about that rather tedious process I decided instead to write a series of stories and poems about living in a Lake District valley.

The book is intended as a tribute to all the hard working residents of the Lake District, and to the wildlife and animals that inhabit the valleys. As a little girl I used to come on holiday to the Lake District and marvel at the natural landscape here.

After living here in a farming community for eight years what has left a lasting impression is the cultural heritage of the Lake District, and the dedication and selflessness of those working here. What inspires me is their contentment with their lives, never wishing for another, easier life and never focusing on material possessions. It can be difficult for a visitor to the Lake District to appreciate this, and I hope that in some small way these stories may illustrate the work that goes behind maintaining the World Heritage Site.

Prologue

I first visited Bethecar Moor in 1993, exploring while my husband was fell running. I found an abandoned farmhouse, and to my amazement the door was unlocked. I let myself in and wandered around the house, imagining who would have lived there in the past. I daydreamed about renting the farmhouse and taking any children that I may have down to Rusland and Satterthwaite School on a quad bike.

I am not the only one to have dreamed this daydream. Talking to several other people over the years in the Rusland and Crake Valleys, they too have had this vision. The daydream took over my imagination, and I wrote the story 'Bethecar Moor' about a character that actually tries to put this dream into practice.

'Bethecar Moor' is fictional. Any resemblance to characters from the Rusland and Crake valleys is coincidental. I hope that the type of characters inhabiting the story, however, are 'bob on' with those actually living here. I hope that my love for and admiration of these characters shines through the story. My thanks to Maria Benjamin for the black and white photographs of Parkamoor.

The short stories and poems are written about my actual experiences living in the Rusland Valley. I hope that I have done the valley and its inhabitants justice.

Certainty

‘There are few things in life
More certain than a hefted flock.
A hefted flock is the very definition
Of certainty, of permanence.’

So you said, washing up
Waving out of the window
At the sheep grazing gently
On the valley sides.

I wish it were true.
I wish our flocks were certain
To continue, but I see them
As the definition of uncertainty.

Nothing is certain in our landscape.
We are fighting a daily battle
Against loss of subsidy,
Loss of grazing, loss of tradition.

We are under friendly fire
From conservation organisations,
Government agencies,
Journalists, environmentalists.

All of them more persuasive
And more verbose than us.
All of them ready to eloquently
State their case and gain support.

Our voices are quiet and calm
But we must make them heard.
Several battles have already been lost
It is time to stand up and be counted.

Cultural Heritage

Holding a sheep at a shepherds' meet
Running in a fell race at a country show
Wrestling in the ring in your long johns
Entering your dog in the waggiest tail class

Baking some cakes for your local show
Buying and selling your own sheep at auction
Watching the hound trials race through the valley
Sleeping out under the stars watching the owls

Driving around in fields in a Land Rover
Catching a fish, tickling a trout
Swimming in rivers, sailing on lakes
Hand shearing a sheep, rolling a fleece

Joining a committee, making a speech
Breeding some puppies, training them up
Helping with hay making, stacking the bales
Eating your lunch in a freshly cut meadow

Mending a wall gap, hanging a gate
Knocking in fence posts, stretching the wire
Putting a broody hen on some eggs
Hatching some chicks, selling the eggs

Judging a sheep show, presenting a prize
Judging a dog show, awarding rosettes
Riding a pony across the fellside
Getting up at dawn for a fell gather

No bucket list required in Cumbria
No list of things to buy to increase status
No retirement, no life of leisure dreamt of
All the life we need is right here in our valley.

The Children of the Lake

As dusk falls the voices travel across the water
Shouts and laughter as they travel home
Birds settle down for the night amid the chatter
Then a torch appears, and the sound of singing.

This is no holiday, no 'Swallows and Amazons'
Adventure, for these are the children of the lake.
Perfectly at home in their native habitat,
Looking through the darkness with clear sight.

After school they come to the boathouse
And rig their boats, then sail off down the lake.
They reluctantly return when it is dark
To hot showers and waiting parents.

Whatever these children do in later life,
Wherever they go, the lake will be in their mind's eye.
At times of stress, boredom or anxiety they
Can internally retreat to the lake.

The children of the lake have an inner strength,
Steeled by winter mornings breaking the ice to sail.
Their resolve is set, their mettle steadfast,
Their belief in themselves is anchored in the lake.

No crisis will be so overwhelming that it
Cannot be solved once the child has learnt to navigate
Deep waters. Sailing through a shifty, changing world
Stormy seas will be conquered by the children of the lake.

Henry's Owl

It was Henry's turn to do his talk to the rest of the class, on a subject of his own choice. I opened the register, scrolled my finger down and said,

'Henry, it's your turn to do your talk today. Would you like to come and do it now?'

'But I forgot it was today Miss.'

'I've been telling you the dates of the talks for two weeks now. I'd like you to come and do it anyway, please.'

This could have been rather a risky strategy, as some children would stand terrified without a thing to say, but I was interested to see how Henry would do.

Henry lifted up the lid of his desk and took out two books that he had recently been studying avidly at every opportunity. They were books about owls. He came and stood at the front of the class, and put the two books down unopened on my desk.

'My talk is about owls' he began 'first I will talk about owls that are native to the UK, then European owls, and lastly some owls that live in other parts of the world such as America.' Henry was seven years old. He spoke confidently and eloquently for the allotted ten minutes, without referring to the two books on the desk. He then answered questions from the class, such as 'Why do you like owls?' and 'What is the biggest owl in the world?' He held the class transfixed. Nobody talked, looked out of the window or asked to go to the toilet. Here was a child born to hold the attention of others, to speak to an audience and think on his feet. There was a rousing round of applause at the end of his talk.

'Despite the fact that Henry had not prepared his talk beforehand, I think that we will give him a mark of ten out of ten. I can see Henry, that if you do not pursue a career as a naturalist, you would be well suited to being an MP, standing at the dispatch box in the House of Commons, answering questions.'

It was quite simply an astonishing performance, not only because of the depth of knowledge about owls, but the delivery of the talk and the ability to answer questions. It was a wonderful thing to be astonished as a class teacher in a small village school, and I wish it happened more often.

I had been astonished on my very first morning at the school. Shortly after 8am my first pupil had arrived, and school did not start until 8.45.

'You must be very keen to come to school,' I said. 'Do you live far from here?'

'Well,' the six year old pupil explained, 'that depends on if you are travelling to school by foot, on a quad or in a larger vehicle.' He proceeded to draw me a very detailed map of his route to school on foot, by quad or pick up, with three separate routes marked from his farm to the village. I knew from that moment, the very first interaction with one of my pupils, that I was going to love teaching here.

I digress, returning to the subject of Henry and his owls; I mentioned the talk to his parents at parents' evening. They told me he had always loved owls. As a younger child he had been convinced that he had one living in his bedroom, and would spend his evenings talking to the owl and refusing to come downstairs. On Christmas Eve his parents had tried to coax him into the sitting room to sit with them and listen to the Nine Lessons and Carols from King's College, Cambridge, but he was too busy talking to the owl.

After the excitement of Christmas the owl seemed to be forgotten, with Henry sometimes mentioning that he used to have an owl living in his bedroom. Gradually he forgot about the owl. It was only when his parents were doing some renovations to his bedroom a couple of years later, when they removed an old Victorian fireplace and the dead owl fell out of the chimney and onto the bedroom floor, that they remembered about the owl.

'I told you,' said Henry.

Joe's Cows

Hill farming always loses its romance in January. It is mid January 2018 and I am home alone on the farm. My husband is away in London working during the week, eldest son back at university, middle son has gone skiing with Young Farmers and youngest son is at school.

I'm in the middle of calving my Dexter cattle. I keep a small herd of six cows, and two have calved already. One of my neighbours is on standby to help me with calving if one should calve while my son is away skiing. It is early morning; the ground is frozen solid, as it has been for days. A heavy fog hangs around in the valley and it feels absolutely freezing all day. The sun doesn't seem to be able to break through the fog.

My neighbour comes past in his Land Rover as I am walking the dogs and puts his window down to ask if any of the Dexters have 'bagged up' yet. Once they begin to get their milk in preparation for the calf we need to be on standby.

'No, I've checked them this morning, none of them are bagging up, so we should be ok today thanks.'

'Just as well, he replied, 'I'm going to be really busy today because our Joe is selling all his cows tomorrow.'

There are few things that people can say to you that actually take your breath away they are so unexpected. I've experienced it a couple of times in my life, once when the

police knocked on my door at 4am, but that's another story. I am absolutely aghast at the news that his brother, who farms in our valley, is selling all his cows.

Joe has not had good health or good luck recently. When helping another farmer testing for tuberculosis he managed to get his hand caught in the unfamiliar cattle crush and sever several of his fingers. Thankfully the operation to re-join the tendons and repair the fingers was a success. I was so overjoyed to see him driving about in his Land Rover again that I ran over to him and gave him a big kiss. This did not go down well with my husband, and I am now banned from kissing him or his brother.

After the heart-stopping incident with the cattle crush, he hurt his back, and decided to sell his cows as another operation and rest was required. There are only a handful of us farming in this valley. It is part of our daily routine seeing each other tearing around in tractors or Land Rovers. Often I pass Joe's yard and stop and say hello to him.

He'll say, 'Where are you going? Don't go far will you?' and when I drive off 'stop again soon won't you?' If I see Joe, or John his brother, it might be the only person that I talk to that day outside of my own home.

I have asked them for help many times, and I'm not ashamed to admit it. For example, once I was out riding my pony on my own in the valley and a piece of stripy tape had been tied to a tree (cycle events often leave this sort of taped route through our valley and the ponies hate the flappy nature of the tape). My pony threw me off and cantered off at speed, standing on my leg in the process making me unable to move, and headed off into the distance still wearing her saddle and all her tack.

Some German tourists on motorbikes eventually found the pony, and Joe got me back up onto my feet and into my Land Rover to go and look for the tack that was scattered around the valley.

'Well I did wonder why you were sitting in the road like that. I thought it was a bit odd.'

Sometimes you need a bit of help and a friendly word. Now, after years of working in the valley, John has knocked the wind out of me by telling me that his brother is selling up. The sale is the next morning and there is thick fog again in the valley. It is so foggy that despite doing the route four times a day to and from school I take a wrong turning on my way to the ferry across Windermere. I miss the hill down to Cunsey from Graythwaite, and end up going through Sawrey instead.

I realise that I am going to pass Joe's brother John's cows. I stop the car and the cows come running expectantly over looking for food. They are Welsh Black cattle, a very hardy beef herd, and are living outside all winter in some woodland. The cows look at me, and I look at them.

'Thank God you are still here,' I say to them. 'Thank you God.'

Meanwhile back in the Rusland Valley all is quiet. Nobody else is buzzing around in a Land Rover because everybody else is at the auction. I drive past Joe's deserted yard, and there is such a feeling of stillness. It is a time of great change in the valley and I feel as if the turf should be ripping itself up from the fields and protesting about the loss of the herd, but nothing protests. All is calm and still.

It is not a good day. Aside from worrying about the auction prices two hoggs, young lambs that have been weaned, have died from pneumonia in the fog despite being vaccinated against it. Four sheep are missing. I spend most of the day looking for them and find two of them eating brambles by a roadside, but I cannot find the other two.

By teatime I am in tears. One or two times a year I will let the emotional turmoil of farming get on top of me, and today with Joe selling his herd is one of them. I decide to have a bath. A warm bath is always a tonic. Unfortunately I have not kept a watch on the Rayburn and refilled it with wood while I have been out looking for the missing sheep. The fire has gone out and there is no hot water.

I fill three enormous old stockpots up with water and

boil them up on the stove. We got the pans from our local Scout group when they were cleaning out a barn, along with a World War One field tent. The pans must be 50 years old.

I carefully carry the steaming pans up to the bathroom, and am about to get into the bath when I get a text from John. The sale has gone well, they are happy with the prices, but Joe was unable to go into the ring and sell his cows himself because his back was so bad. 'Job's totally knackered,' says the text, meaning everything is broken. He's not wrong.

Winter in the Valley

When we bought our land at Blawith in the Crake Valley it came with a resident claiming squatter's rights, a very wild Swaledale ewe. She did not belong to the previous owner, and it proved very difficult to catch her and find out whom she did belong to.

One day my youngest son became determined to catch her. If only we could contain her we could look at her ear tag number and at least find out where she was born if not who she belonged to. After about half an hour of careful maneuvering we managed to trap her in a corner next to a cattle grid. It would be useless sending one of the dogs to 'hold her up' as she had no respect for man or dog and would head barge.

We stood eyeball to eyeball for about thirty seconds, the sheep with her back to the cattle grid, and my son and I blocking the exit. In one swift movement that looked to me as if it had been choreographed for the Royal Ballet, my son and the sheep simultaneously leapt in the air. I watched and it appeared to be in slow motion, as my son leapt higher than the sheep, reached down and put his shepherd's crook around her neck whilst she was in mid-air, and then both landed on the tarmac with him sat astride the sheep with his crook around her neck. Realising she was caught, she now began to chew her cud, as calmly as you like.

It was the most remarkable piece of sheep catching

choreography, expertly implemented, without either of them getting hurt. Quite remarkable. He sat astride her and said, 'Told you I'd catch her.'

We now had her ear tag number and could phone and find out where she had been born, which turned out to be about 150 miles away. Someone must have bought her at an auction, so we put her in the back of the Land rover and did a little tour of surrounding farms trying to persuade someone that she must be their sheep, or at least persuade somebody to keep her.

No luck, nobody wanted her. At least we had tried to return her to her rightful owner. It didn't seem morally right to sell her at auction when she was not our property, and I dreaded the owner turning up to claim her and telling him she had been sold, so there was no solution other than to re release her back onto our land.

She mixed well with the Rough Fell sheep when they arrived, and was always easy to spot in a crowd as she was the only Swaledale. She ran with the Rough Fell tup each year, and produced a steady run of Rough Fell/Swaledale daughters to join the flock.

By the time we had had the land for four years she was obviously getting on in age. After shearing all the sheep are sorted through each year. Those with no teeth are unlikely to survive in our harsh environment, as they cannot get enough nutrition from the poor grass once they are 'broken mouthed'. I was dreading 'Old Swalie Lass' going down the race towards my son who was making the decision about which ewes had to go.

He opened her mouth, closed it and said, 'As you well know she has no teeth. She should go. She'll not survive another winter, but I know you're not going to sell her. Just to let you know, she's going to eat a load of your cake (sheep food) and then drop dead.'

To my relief he let her through into the pens where the sheep to stay were. I knew she was probably going to die, but I thought she deserved to die at home.

Every morning as I drove over to the field where she was living in winter there was a rush of sheep. They would all wait at the top of the field and then run down the lane baaing once the Land Rover appeared. All that is, except Old Swalie Lass. She was too canny to expend energy running. She was already standing waiting at the bottom gate each morning because she knew that the food was going to come through that gate.

She was the last to leave the feeding troughs each day, making sure she got every last bit of food. She sucked every last bite up like a vacuum cleaner going along the bottom of the galvanized troughs.

It was our last week in the valley. We were all anxiously waiting for the purchase of our farm to go through so that all the animals could be moved. My son asked if Old Swalie Lass would be going or staying, but I had no answer to that. She wasn't really my property to move.

It was a cold Tuesday morning, the kind of morning when you just have to steel yourself to get on with it. I had two layers of thermals on underneath my clothes and waterproofs on top. When I arrived at the top gate all the sheep

were shouting at me, telling me how cold it was. I glanced down the field, and there was no Swalie Lass. 'She's gone' I thought. I expected to find her body frozen next to a dry stone wall, but there was no sign of her. I fed the sheep and began a search. She was not in any of the sheep fields.

The ponies are involved in the process of rush removal, conservation grazing to rejuvenate a field that has become over run with rush. It is a very boggy field, totally fenced off with new fencing so that the ponies cannot wander off to find better grass.

As I fed the ponies their hay and carrots I spotted something in the middle of the bog - Old Swalie Lass! How she had got into that field over the fencing goodness only knows. I began to walk over to her, but it was extremely boggy and I had to step on clumps of rush to keep on top of the water.

I got to within about three meters of her, and was trying to see if she was alive, when I lost my footing and both legs went into the bog. Water filled my wellies, and somehow a vacuum was created so that I was stuck, thigh deep, in absolutely icy water.

I could see now that Old Swalie Lass was dead, but there was no way that I could reach her safely. I began to worry that I too would be stuck in the bog. I lay down across some clumps of reeds and pulled myself out of the bog with a squelch. I was then able to walk back to the edge of the field. I was absolutely soaking, and totally covered in mud. It was several degrees below zero, and the wind was blowing relentlessly.

I turned and said goodbye to Old Swalie Lass, and then drove home with chattering teeth. The Rayburn was not on, and there was no hot water. My husband was at home, and helped me pull off my waterlogged wellies and boil some pans of water on the stove to have a wash. At times like these my existence seemed almost medieval.

By the following day there was no sign of Old Swalie Lass, her body had been taken down into the bog and there was no way I could retrieve it for the fallen stock man. She must have known change was afoot. She hadn't wanted to leave that land, and now she didn't have to.

Winter seemed endless. Relentlessly cold and wet. No dramatic snowfall, just day after day of wind and rain. By January we have lost three residents of the valley. My son says, 'Well it's winter Mother, people die. I have been to fourteen funerals in my life, no weddings and no christenings.'

This is not strictly true as he has been to both a wedding and a christening as a baby, but was too young to remember them. He is right that his main experience of rites of passage in this valley has been through funerals.

The day after the death of Old Swalie Lass is the day of the funeral of a much loved neighbour. I often think that the weather in our valley reflects the mood of the inhabitants, as well as the inhabitants' moods being influenced by the weather. It is a bitterly cold day again. The funeral is at 12.00, but by 10.30 there is a steady stream of cars along the valley heading for the church. It is not a small church, but by the time I get there at 11.30 it is standing room only.

There is only one place left to stand, right next to the choir. The choir mistress jokes, 'Are you singing in the choir?' to which I reply 'I haven't sung in a choir for about

30 years.'

'Can you read music?' she asks, and before I know it I have been co-opted into the choir.

The graveyard outside fills up with people standing. When there is a funeral from a farming family it is not uncommon for people to have to stand outside, but today it is so bitterly cold. As well as family, friends and neighbours, the staff from the auction marts and feed suppliers are in attendance. Everyone is frozen.

At 12.00 exactly the funeral cars draw up outside, the Vicar walks across the graveyard to meet them at the gate, and there is the most almighty hailstorm. The Vicar's hair and shoulders are covered in hail. It is a spectacular weather event, it seems almost biblical. The weather is reflecting the mood of the valley's inhabitants again.

The following day I speak to my neighbour about the funeral. I say that I couldn't believe the ferocity of the hailstorm, but he tells me that he could believe it. When our neighbours were married there was also a hailstorm like that, and their farm yard was filled with hail that evening. Then the night that she died the same happened, the farmyard was filled with a carpet of hail.

'It's like the weather is sending you a message,' I pondered.

'But what would the message be?' my neighbour wondered.

'That she's home safe?' I suggest.

'Maybe, maybe, anyway enough chatting I've got jobs to do,' and the Land Rover winds it way down the valley and into the woods.

Animals and people, deeply connected to the landscape

and trying to survive whatever weather we are sent. We work with the weather; it is pointless to fight it. 'It's winter, Mother, people die.' So they do son, so they do, and generally in this valley we accept that and just keep on working until nature comes to claim us for its own, and we return to nature in the graveyard in the valley. There is no real retirement here. You work alongside nature until one winter proves to be too long.

Valley Legacy

What I like about
Our valley
Is its self-sufficiency,
Its ‘valley-sufficiency’
Regardless of the self.

We always try
To sort things out
Amongst ourselves,
Calling in outside help
Is a last resort.

When one of us dies
The body is kept
At home, safe,
Then ratchet strapped
Onto a bale trailer.

Driven to the funeral
On a favourite tractor
Cleaned and polished
Family following behind
In pickups and cars.

You might have a
Local event or trophy
Named after you.
A tractor run
Or a fell race.

That is your legacy.
But legacy is also
That you live on
In other's work
In the valley.

When I drive past Henry's gate
And his nephew is
There with the cows,
It is Henry I see in
My mind, every day.

Henry's legacy is
His family farming,
But it is also Henry
By the gate waving
In my mind's eye.

Twice a day
I see Henry.
As time passes
I will see more old friends
Only in my memory.

Lost in Lichen

On a gate post or a wall top
There is a miniature world
A forest of lichens and mosses
The names of which I do not know.

Lately I have begun stopping
And looking into these landscapes,
Losing myself in the vegetation,
Wandering visually thro' the undergrowth.

There is another world here,
Another layer to our valley,
That I did not know about.
Most people pass by without a glance.

This world is stunningly beautiful,
Particularly when preserved in frost.
Frozen jewels glinting in the sunshine
Of a midwinter's day.

There is more to this valley
Than meets the eye at first glance.
I now stand and look a while
Lingering amongst the lichen.

Valley of the Owls

By day people think
They are in charge of this valley
But go out after dark and you will see
It is really the valley of the owls.

In the morning
As I walk my dogs under the pop hole
I can hear the barn owls snoring
Biding their time until dusk.

At dusk, or just before,
The barn owl swoops out of the barn
And glides down over the meadows
In the valley bottom searching for food.

Such a skilled hunter
Is he that we have had to create
Dormice corridors to allow the mice
To travel about safely without being eaten.

Barn owls then roost
In the trees behind our house
And shriek to each other in delight
Throughout the night. The night is theirs.

VALLEY OF THE OWLS

Tawny owls follow barn owls
Slightly later to rise
They wait until it is properly dark
And then begin calling to each other.

The valley echoes
On all sides from owl calls.
We may think we own this valley
But by night it is the realm of the owls.

The Bridge over the Stream

'I don't know how many more years
I'll be bringing me cows this way
So I thought I'd get a photo
While it's a sunny day.

'You just stand on yon side
And I'll walk t'cows ower t'bridge
Mind you get 'em all in though
And t'fells behind along t'ridge.

'I'm going to take the photo
To an artist up in Dale Park
And get him to paint me a picture
Of me and me cows up with t'lark

'And when I'm an old man in Rusland
Sitting in front of the TV screen
If I look up above me hearth
I'll see me cows on't bridge ower't stream.'

Celebrity Lengthsman

We don't have many celebrities in our valley
But we do have a 'celebrity lengthsman'.
What is a lengthsman? He is paid to 'walk the length
Of the parish keeping the ditches and drains clear.'

If you have a blocked drain or a flooded road
The lengthsman will come and sort it out.
He is an invaluable emergency service in
Stormy weather. Digging out the drains.

What makes our lengthsman a celebrity
Is that his drain spotting has 'gone viral'.
Hundreds of people have bought his calendars
Showing a variety of traditional parish drains.

Spurred on by his success he has also
Produced a milk churn stands calendar
Showing the variety of old stands around the parish
TV, radio and newspapers love his smiling face.

Day by day the celebrity lengthsman
Works hard on the roads with his warning triangle
But in the evening you will find him after dinner
speaking
About the beauty and idiosyncratic nature of our
parish drains.

In Scruff

Early morning, darkest winter
Passing by the shippon in the car
Raymond comes out into the daylight
Blinking like I imagine a mole would
On breaking through the turf.

He is wearing his overall,
And an old jumper, both dusted in straw.
A wool hat is set on his head at an odd angle,
Also covered in straw,
He hasn't shaved and rubs his face.

Quickly I take a photo
As I lean out of the car window.
'What are you going to do with that?'
He asks 'you young ones, tweeting,
I don't want that going viral.'

'This is just for me' I reply
'I'm going to keep it on my phone,
And when I've moved away and
Am in need of inspiration I'm going
To get it out and look at it.

'What I have here is a photo
Of someone happy in their work,
Completely connected to their land,
Getting on with their life
And at peace with themselves'

‘Well’ he says, ‘if that’s what
You’re going to do with it, I don’t
Suppose I can grumble about that’
And he walks off rubbing his beard
And laughing to himself.

Another day I get a second photo
Of him passing by the house on a quad,
Two border collies riding shotgun
And a pile of yellow buckets
Tied on with bailer twine.

‘And what are you doing with
That one?’ he asks ‘another
Photo of me in me scruff’
‘I’m selling it to the *Daily Mail*’
‘Aye’ he laughs, ‘you bloody would!’

Hulleter Sheepwash

When our neighbour Alan died
There was a stillness
In the valley.
Not only with the loss
Of our beloved farmer
But because there were
Suddenly no cows.

Every evening school run
We would be stopped
At Hulleter sheepwash
By the slow march
Of 120 dairy cows
Making their way home
For milking.

We would hang out
Of the car window
And exchange a few
Words with our neighbour
On his 'motorbike'
Always cheerful and happy
Content with his life.

HULLETER SHEEPWASH

When he died
On his farm, by his fire
In his armchair,
Like his brother had died
In the milking parlour
A few years before,
The valley fell silent.

The cows were sold
Off to new homes
And the fields stood empty
Waiting for something
To happen.
No smoke from the chimney.
No barking dogs.

Local people stopped
Their cars and stared
At the empty farm.
Bereaved.
Life in the valley was changing.
He was buried in
The graveyard by our house.

My youngest son
Would often disappear.
'Where have you been?'
'Talking to Alan'
'How is he?'
'Dead, Mother,
He's dead.'

A big sign was put up
'Farm for sale
By Private Treaty'
But nobody seemed
To look around.
Nothing happened
For months.

Then unfamiliar
People arrived to
Work on the farm.
Repairing walls
Putting in gateways
Working for
The new owners.

The sign came down
And sheep arrived.
At last the secret
Was out. A local family
Had bought the farm.
Smoke appeared from
The chimney again.

If you are passing
Hulleter sheepwash,
Slow down, you might
Be met by 100 cows coming
Towards you, not dairy cows
But a beef suckler herd.
Enjoy the view.

If you pass the farm you
Might see someone working.
Even late into the night.
Working hard to
Improve the land.
Neighbour Alan
Would be so proud.

Things change in
The valley. People
Come and go.
But thankfully for now
The farm continues.
The valley has its
Heart and soul intact.

BETHECAR MOOR

A Lakeland Story by Andrea Meanwell

1

Sawrey:
There was never one defining moment, there was never a straw that broke the camel's back, more a growing feeling that it was time to leave the family farm and strike out on my own. It was my time now, and I had to act.

I had packed a rucksack and two panniers for the pony with the provisions that I thought I would need for the first week, and silently slipped downstairs with them into the farm kitchen. The kitchen was always cosy because the Rayburn stove burned 24 hours a day, 365 days a year. It was considered unlucky to let a farm kitchen fire go out. I was going to somewhere where the fire had been lit for about 500 years, but at the moment the kitchen would be cold and empty, standing lost in time on a Lake District fellside.

The emptiness of the farm had been annoying me for several years now. A traditional Lakeland farmhouse and barn, abandoned by the conservation organisation that owned it, as they did not consider it was suitable for 'modern living'. There was no road to the farm, no electricity, no running water and no mobile phone signal. I intended to prove that it was possible to live a 'modern' existence there. I intended to occupy and repopulate the farm, to breathe life back into the moorland there by managing the landscape. Nature abhors a vacuum, and the fields were filling up with rushes and brambles. I was going to bring the farm back to life.

I had to be quiet as I crept through the kitchen, mother and father would never approve of my plan to illegally occupy a farm. I left them a note saying simply ‘Gone to Parkamoor’. They knew that I liked to visit the farmhouse up there, and had slept there before. They could find out at a later date just how long I intended to stay at Parkamoor.

I whispered to my little dog Bee, and she skittered from her bed and stretched, ready to go wherever we were going, and we silently slipped out of the kitchen and into the moonlit night. Without a clear evening the plan would have had to be put off, as we needed the light of the moon to find our way.

Apple the pony was waiting in the stable, ears pricked as we opened the door, blowing through her nose as the cold air from outside rushed into the stable causing her to snort. I took Apple’s tack down from the hooks on the wall. She always wore a headcollar, as this made her understand that she was ‘caught’ and not running wild on the fells as she had as a foal. Her bridle was put on over her headcollar, and a martingale to stop her from pulling too much along with her saddle. The panniers were attached onto the saddle. The saddle itself sat on a soft numnah, so the extra weight wouldn’t chaff her skin. Her girth was tightened, stirrups lengthened, and she was walked outside to the mounting block.

Apple never questioned why we were going for a ride, and the dark did not seem to worry her. Rucksack on my back, I gently slid onto the saddle and told Apple to walk on. Bee followed behind her swishing tail at a safe distance.

Our farmhouse was set at the head of a valley in the 'V' of the valley bottom. The moonlight reflected on the white paint, making it luminous in the dark. Quietly we walked past the house, right under the bedroom window where my parents were sleeping. At least mother would not be able to ask me 'What are you going to do now that you have finished college?' over breakfast, because I would be actually doing it.

We walked around the corner on the road, and then down the path to The Strands. The river was flowing fast, but was not too deep and Apple could see where to put her feet as the moonlight was lighting up the water. She knew that there was a large stone to the left of the footpath, and if she stepped right next to it they could easily walk across the river without slipping.

Once on the other side we waited and called for Bee. Poor Bee did not look very amused at having to swim through the river during the night, but she jumped in and swam across. After she had shaken herself we began our journey again. We had to travel through Grizedale Forest; a large Forestry Commission owned wood. Apple and I were very familiar with the route along the forest tracks, and she now sensed where we were going.

I don't know if you have ever slept on a pony's back, but it is very easy to fall asleep and have a little snooze while you are riding a trusted pony on a familiar route. I often close my eyes and have a little nap on a well-known route. Tonight, however, was different. All of our senses were very alert as we were not used to riding at night and there were unfamiliar sounds in the forest that seemed to echo around us.

After about an hour we came out onto the open moor top of Bethecar Moor. Apple knew that there was now a long stretch of grass and this was her opportunity to canter, she increased her pace in anticipation of me telling her she could canter. 'OK girl' I said, and off she went. Apple's mane stretched out in the moonlight as she gathered speed, and as tiny Bee sprinted along trying to keep up with her ghostly shadow, we made an unconventional scene as we travelled along the moor top. It is always exhilarating to canter along a fell top or over a mountain pass, but tonight we all sensed that we were at the beginning of a great adventure.

As we approached the wall across the moor Parkamoor farmhouse came into view. An abandoned farm on a Lakeland fell top that we were about to make our own. Soon the fire would be lit and the shutters would be opened in the morning to let the daylight in for the first time in years. It was to be the beginning of a new chapter for Parkamoor, and for one girl, her dog and her pony.

The farm at Parkamoor is situated at 600 feet above Coniston Water on Bethecar Moor. The monks of Furness Abbey established it in the fourteenth century. The last permanent resident died there in a snow storm in the 1940s, and since then the farm has been used as a bunk barn, artist's retreat and as a holiday let. From its incarnation as an off grid holiday let the farmhouse was still fully furnished. It was nearly impossible to get furniture up there, never mind remove it. I had stayed at the holiday let several times, and knew where the key box was located and the code for the key. Due to its remote location, and lack of electricity and running water, only the most intrepid of

holidaymakers had stayed there, and the house now stood empty.

As we rode up to the house the moonlight reflected off the windows, and it was easy enough to see the combination lock for the key box and enter the code for the key without a torch. I jumped down from Apple, took off her panniers and riding tack and let her loose in the small paddock next to the house. She stood there watching me, wondering what I was going to do next. That was the big question in my life really, what was I going to do next?

The cold air focussed my thoughts, and I opened the

front door and stepped inside. Bee waited to be asked to follow me, and then shot over the threshold and into the parlour. I carefully stored all of Apple's tack at the side of the hall, next to the large slipper bath that was sitting there.

The main room in the house was the parlour. In here there was a large black range, with a water tank and oven next to the fire grate. The most urgent job was to light the fire and fill up the water tank. I pushed through the woollen curtain to the left of the range and into the storeroom, and loaded up my arms with wood and kindling. Thankfully there was a good supply of wood in the house. I had brought a box of matches in my rucksack, and soon had the start of a fire in the range.

I then went back into the storeroom and fetched some water from the old pump in a big kettle and began to fill the water tank next to the fire. I would have to do several trips to fill it, and after following me for the first few trips Bee settled herself down on the hearth rug content to sleep in front of the fire.

After about an hour I had a full tank of water and a blazing fire, so I turned my attention to light and lit some of the candles that I had brought with me. I had a small 'wee Willie Winkie' candleholder, and put one of the candles in this to have a look around the house. As well as the parlour there was a small sitting room downstairs, and a small kitchen/pantry. A kitchen without running water and electricity was going to be a challenge, but at least there was somewhere to chop and prepare food.

Quietly Bee and I took the candle upstairs and stood in the doorway of each of the three bedrooms, looking in. All the beds were made up, no doubt in expectation of some holiday guests who had never arrived.

There was no bathroom, the slipper bath would have to be filled in front of the range, and there was an outdoor privy. Ash from the fire had to be put into the privy each morning to help with the composting process.

Tomorrow we would get to grips with day to day living at Parkamoor, but now it was time to sleep. I unrolled my sleeping bag from my rucksack and spread it out on the floor in front of the fire, and climbed in. The fire was lit, the water was warming, the farmhouse was occupied, and tomorrow I would begin to bring the farm back to life.

2

Day to day living at Parkamoor was really just a series of jobs that had to be done. Firstly and most importantly the fire had to be kept burning. Thankfully there was about four years worth of wood stacked up around the back of the house, although I was aware that I would have to pay somebody for it at some point. Also at the back of my mind was that somebody at some point would come and challenge my right to be here. This would probably be the farmer who leased the land from a conservation organisation, as this old house was included in the multi-generational tenancy of his farm.

When the time came I would appeal for leniency, and ask for permission to stay for the winter. There would be little chance of them getting any holiday guests up here in the winter, so they may as well have me as caretaker over the winter months. I had to hope it was a persuasive argument.

In the morning the cold ash from the bottom of the range was taken outside and put into the privy. You had to be careful when you opened the door as a gust of wind in the wrong direction could easily blow the ash into your face and hair. The fire in the range would then be stirred up for the day, and the water tank filled.

Small quantities of water for washing could be carried through the house and into the kitchen, but for washing larger items such as the bed sheets I found it was easier to wash them outside in the stream. They would dry in the wind pegged out in the farmhouse garden. I had washed all the bed linen in the house, although there was no way to iron it. I slept in a small single bed, as it was easier to wash and dry the bedding.

Food, candles and any other provisions had to be bought in the village on my once a week trip there on Apple. Her panniers would be filled up. I had enough money to live frugally for the winter, saved up from part time jobs while I was at agricultural college. I would be unable to live like this forever.

On my way down to the village I noticed that another

farm, Lawson Park, had a letterbox at the lakeside so that the postman did not have to travel up the hillside to the farm. I found a small wooden box with a hinged lid in the storeroom, and painted Parkamoor Post on the front, and placed it at the bottom of the path down to the lake. It was a bit like *Swallows and Amazons*, waiting for someone to send me a signal.

On my next trip into the village I saw our village postman Bob Satterthwaite, and flagged him down and told him about the box. He told me not to be expecting many letters from my parents as my mother was in a 'terrible fury' and was expecting me back any day from my ridiculous escapade. Such comments were likely to make me much more determined to stick it out.

It was a physical life of chores. Wood, water, washing, sweeping, cleaning, fetching food and provisions, but it was one that I enjoyed. I was never bored as there was always plenty to do, and I was always exhausted at the end of the day and slept soundly.

After two weeks of checking an empty letterbox I found a note from my mother asking when I was going to come home. I sent one back when I was in the village, saying I was intending to stay for the winter. My mobile phone had run of credit and I had no way of charging it, there was no mobile signal anyway so all messages had to be sent by post.

In the afternoons I usually walked Bee along the moor top, or took Apple for a ride with Bee trotting behind. It was a blissful, peaceful existence, but there was a definite nip in the air and winter was approaching.

Autumn is a very short season in the Lake District, and

before long winter had us in its grip. Frosty mornings and sunny afternoons if we were lucky, endless rain and wind if we were not. Dashing out to the outside privy was not fun in the wind and rain, and I sometimes questioned what I was doing there.

I was living in someone else's house, using someone else's wood, three miles from the nearest road, without any method of supporting myself in the future. I would have to do better than this. If I couldn't justify my existence up there to myself, how was I going to convince the owner of the property and the tenant farmer who leased it that I should be there? I would have to get a grip, and get together some sort of strategy for surviving the winter. It was all very well to collapse into bed exhausted every night and sleep soundly, but what was my plan for when my money ran out?

3

Cold frosty mornings with clear skies really bring out the best of the Lake District. From the tiny window in my outdoor privy I could see down the hillside and along Coniston Water. It had to be one of the best views from a toilet in the world, I thought to myself each morning. On some days you couldn't see the lake as it was shrouded in mist, but on other days you could make out the small white sails as early morning sailors tacked across the lake and around Piel Island.

On my way back to the house from the privy one morning I encountered some visitors. Fifteen Herdwick sheep came barrelling around the corner of the house and stopped dead still when they saw me. Straight away, without looking at their smit mark or ear tags, I recognised them as my father's 'rogue' sheep that had gone missing about four months previously. We had spent many hours searching along the valley bottom and through the woods for these sheep, but of course being Herdwicks they had headed for the hill top.

I picked up an old feedbag that lay discarded on the floor and rattled it. The sheep seemed interested. Rattling the bag with them following me I managed to get them into the small paddock with the very high walls next to the house where Apple was grazing.

I fastened them in, and my mind started whirring into

action. This could be my starter flock; I could heft these Herdwicks to the fell next to the farmhouse, once I had tamed them in the paddock. My father had long since given up any hope of finding them. These could be my sheep, my future.

After their wanderings the sheep seemed surprisingly content to stay quietly in the field. They could be contrary like that, Herdwicks; one minute they were as calm and placid as a downland sheep, the next they were ratching over walls.

I decided to write to the conservation organisation. Having the sheep here seemed to make my existence more plausible and more sustainable. I wrote them a long, rambling letter about hefting the sheep, and myself, to the fell. I posted it in the village. I checked my post box for a reply, but none was forthcoming. I began to wonder if my letter had been lost amongst a mountain of paperwork on somebody's desk.

The reply came not in the form of a letter, but a visit. One clear morning I was washing some clothes in the stream close to the farmhouse when Bee started barking to warn me that someone was near. Then I heard the unmistakable sound of a quad bike. The farmer, who I recognised from the farm down the hill in the Crake Valley, drove alongside me and switched the engine off. You always knew when it was going to be a long conversation with a farmer, as Land Rovers and quad bikes are very rarely turned off whilst chatting. He sat for what seemed like an eternity on the quad, watching me, as if preparing his ideas on the situation. Eventually he strode off the quad bike and walked slowly towards me.

'Sawrey Moor?'

'Hello.'

'Do you enjoy making work for yourself, washing in that stream? Why don't you go home and make life easier for yourself?'

There followed a conversation in which I tried to explain what I was doing, he kept taking his hat off and holding it in his hands, and then putting it back on again, as if he was unsure about what to do.

'I've had a letter from me landlord that says its up to me whether you can live here for the winter, and I must decide, like,' he puzzled. There was a silence between us, during which the wind blew and buffeted my hair across my face, and I tried to predict what he was going to say next.

'You can't stay without any hay for them ratching Herdwicks.'

'Do you have hay I could buy?'

'I might, lass, I might.'

I knew then that things were going to be alright, and that I was going to be able to stay. My heart sang as he told me he would deliver me my first five bales of hay tomorrow, as bad weather was forecast.

The next day the wind blew in a way in which I could tell that bad weather was coming. In the early afternoon the grey clouds chased each other across the sky, followed by snow clouds. I paced backwards and forwards in front of the house wondering if he had forgotten about the hay.

Just before it began to go dark I heard the sound of the quad bike, and it came along to the house with a small quad trailer loaded high with hay. I stood in the porch and watched the bike and trailer struggle through the snow. It stopped and I ran out to help unload the hay. We would have to move quickly to get it into the small barn next to the house before it got wet.

'The last resident here died in a snowstorm,' he cheerily told me.

'Come inside and tell me about it,' I said.

The hay was unloaded, some thrown to the sheep and Apple, and we hurried inside to the range and kettle. Both Bee and his dog Fly settled down by the fire.

He told me about the last family to live here, and how he had often wished that a family lived in the house now. He told me about his farm in the valley, and how he had no children to take it on. The snow was falling thick and fast inside. We went to the porch and opened the door.

'I don't think you can go anywhere tonight,' I said

'Sawrey, what makes you think I want to go anywhere?' He said. He lifted my face up towards his and kissed me. This took me completely by surprise, as did my reaction in kissing him back, and we both found that tenderness and affection were a cosy way to block out the snowstorm and sleep soundly that night.

The next morning I went down to stir up the fire, and two dogs lifted their heads up from the hearthrug. The snowstorm was over, and Fly and Bee were keen to get up and out. My visitor came downstairs, put on his hat and went outside to start the quad. As he drove off through the rain I wondered how he would explain his absence to his family, and to himself.

4

After that encounter I left money for the hay in the barn, and the hay appeared while I was out. I was aware that he must be watching my movements, but didn't actually set eyes on the him again. I concluded that he could not trust himself to visit again, and got on with the day-to-day chores of fell farm living.

Living on a fell farm in winter is really just a series of jobs to do. Keeping the fire burning and the water warm, keeping the animals fed and watered. There was no stream in the field where Apple and the sheep were living so I had to carry buckets of water to them, and break the ice on the top of the buckets several times a day if it was cold.

I decided that the best time of year to heft the Herdwicks would be in winter, as in summer the fresh grass would be too tempting if they were newly released onto the fell. I began to feed them at night, and let them out onto the fell-side in the morning. They would wander around by the house in the daytime, and I would shut them into the paddock with hay in the evening. Similarly Apple began to wander onto the fell in the daytime. I knew that she would come back as there was no way that she would resist the lure of hay.

Evenings were long and dark without electricity. I began knitting by candlelight and made some hats to sell in the 'Honest shop' in the village. This was a small unmanned shop in which locals could sell their products to visitors and other local people.

I got into the habit of sending my parents a postcard every week with a quick note about what I was doing. I told them I was hefting the Herdwicks, and not to worry about me. As time went on it became apparent that the Herdwicks must have met a male suitor on the fell at some point as they began to become heavier and were obviously in lamb. Tups, or male sheep, should not really be released onto an area of shared grazing such as a common, but sometimes an uncastrated male lamb from the previous year, or an escapologist tup, can find his way to the females.

I had a lot of sympathy for their expanding size myself, as despite my frugal existence I too seemed to be putting on weight. There was a rosiness to my cheeks that showed that the mountain air was agreeing with me. I was settled, happy in my routines, content with my life. I tried not to think too far ahead.

Spring would come, as surely as winter had followed autumn, and it was just a question of keeping on top of everything until the better weather lifted our spirits and the sound of birdsong woke us in the morning announcing the arrival of spring.

Every little sign of spring was eagerly noted, and the next sign anticipated, on the hill farm. First came the snowdrops, then the small wild daffodils, then the mosses and lichens, and finally the spring grass started to grow after the primroses and around the time when the bluebells were in flower.

Once the grass began to grow my life became easier. Hay didn't have to be carried to the sheep. The sheep had to be contained now spring was here. I dare not let them onto the fell in case they got 'spring fever' which is a kind of 'grass is always greener' mentality where they travel from field to field, forest to common, devouring fresh grass in a frenzy. The arrival of spring can turn us all a little crazy.

Spring certainly affected me as a hill farmer; it put a smile on my face and a spring in my step. I was woken at dawn every day by birds singing in the hawthorn bushes, and before long the first lamb arrived in the paddock. I had no need to assist the sheep with lambing, Herdwicks are a native breed that is exceptionally hardy and can usually lamb themselves.

I was delighted to see that the mystery gentleman they had been seduced by on their travels was also a Herdwick, and the lambs were pure bred. After seventeen days each sheep had a lamb, and I had eight gimmer lambs that could join my flock and increase my numbers.

I loved to sit in the evenings on the small hill top near the house and watch the lambs play, running about in lamb gangs and tiring themselves out before bed.

No notice to quit had been received from the farmer, so I kept myself to myself, living quietly and tending my flock. I did not allow myself to think of the possibility that I might have to go home. Spring was turning into summer, with long balmy evenings watching the swallows swoop around the farm. Unlike the swallows with their exhausting migration, I had begun to feel hefted. I had no desire to fly. I was grounded.

5

May was hot and heavy. Humid air sat on the fell top and did not move. I was hot and heavy and did not feel much like moving. We often get our hottest weather in Cumbria in May, and it had come now encompassing the Furness fells in a heat haze.

There seemed to be no end to the hot weather, which lasted all through May and into June. Thunderstorms provided the only respite from the muggy weather, refilling the tarns and streams and thankfully saving us from drought.

The lambs and their mothers lived happily in the paddock, and I was content to stay at home and watch them play. Apple sheltered from the intense heat of the sun by standing under an oak tree in the paddock swishing her tail to bat away flies. I did not feel like going anywhere or doing anything. I had not been to the village for two weeks.

Despite the hot weather jobs still had to be done, and one morning as I was washing some clothes in the stream I felt a hot burst of water between my legs.

'Oh no, not now. I am not ready," I said. Then I shouted, 'I am all alone!' At that Bee, unaccustomed to me shouting, shot off across the fellside.

'Bee, Bee' I shouted, but was stopped by a searing pain filling the bottom of my stomach. Between pains I worked quickly. I felt like I needed to be in the bath, in warm water to ease the pain, so I dragged the slipper bath in front of the range and began to fill it from the copper kettle.

I had to stop with each pain, but I continued to fill the bath. All farmers are used to working with pain. There are no days off in farming. Even if you have a stomach bug and are physically vomiting, you have to do it between jobs.

I tried to remain calm, calm and quiet. 'It is a good pain. Good will come of it,' I told myself over and over. I undressed and slipped into the bath. Instantly I felt better, and half nodded off despite the pain. My body seemed to be taking over, my mind wandered off.

Some time later I thought I must be delirious, as I seemed to have my mother by my side. Between pains she was telling me a story about not getting postcards, coming to see if I was alright and being met half way by Bee barking. She had run the second half of the way here, and was fussing about looking for clean towels.

Then she was holding my hand and telling me to push. I pushed as hard as I could, and then in my strange dream like state I saw the face of my child swimming up through the bathwater towards me. Everything seemed to be in slow motion, and I saw her face and two little hands making swimming movements clearly underwater for what seemed like hours before my mother placed her on my chest.

'Well done. Well done,' my mother said. I thought she was talking about me giving birth, but then she went on "you've solved the succession crisis at the farm in the Crake Valley.' My mother was gazing down at my newborn daughter, still attached by her umbilical cord.

Less than one minute old and already heir to one of the

best farms in South Cumbria. I looked at my mother, lost in thought, and thought what a waste of time it had been wondering how I was going to tell her that I was pregnant.

'People talk, Sawrey. People talk,' she muttered, by way of explanation.

Then I realised in a moment of clarity that even on the most remote hill farm I was still connected to society, still part of it, and my daughter and I would have to become part of local life if we were to be accepted.

'I'll call her Bethecar, Bethecar Moor, after where she was bor,' I said.

'Aye, and where she'll farm,' said my mother. Such a tiny girl and such a weight of expectation for her future.

Somehow, by trying to break free like a fraying strand from a piece of woven fabric, I had become even more interwoven into the tweed fabric of our valley life. That was never the intention, but there was no escaping it now. My daughter and I were part of the fabric of the cultural heritage and succession of farms in the valleys surrounding Bethecar Moor, and would always be so. The umbilical cord that still tied us together also tied us to the farms and to our future here.

6

Ruth:

It is difficult to know where to begin, so I will begin in the most obvious place, in the beginning. I was born in my parents' bedroom on Christmas day 1970, while the wind rattled around the house like a hooligan. There were strikes and power cuts all over the country, but we did not have electricity yet at the farm so that did not trouble us.

My parents were tenant farmers at High Bethecar Farm, which is perched on the side of Bethecar Moor overlooking Coniston Water. Positioned out of the wind, the farm buildings tucked into a sheltered spot, but the wind on my birthday was coming from the west and straight onto, or through, the house.

I was my parents' first child, and was joined by my brother Robert twelve months later. Together we lived a wild, unfettered life on the farm. I don't know how it was decided which farms were to have tarmac roads put up to them, and which did not, but to this day there is no 'proper' road to the farm just two tyre tracks along the edge of the fields. You would never have imagined that anything lay at the end of the tyre tracks, but our whole world was there in what would today be called an 'off grid' existence.

There were plenty of places to play, like the old farm buildings at Low Bethecar where I set up a mini house with a table and chairs made out of old boxes to entertain my dolls to a tea party. We both helped with jobs on the farm, particularly at lambing time when we fed the orphan lambs their milk and ran about in the yard with them. I had a little terrier called Binks who accompanied me everywhere.

It was a happy, carefree existence for us children, unaware to a large extent of the difficulties in making a living hill farming. We were totally immersed in our landscape, climbing on hay bales, lying in meadows in the sun watching dragonflies, catching ladybirds in jars.

The only blot on the horizon, more so for Robert who hated being indoors than for me, was school. When I was five years old I had to attend our village school. It was about four miles to the school. I had to walk down to the edge of our farm, about a mile, then a minibus picked me up to take me to school. My mother and Binks walked with me, and when I got off the minibus in the afternoon Binks would be waiting to walk home with me.

I enjoyed school; there were lots of children from local farms there to play with. If we had been up in the night for any reason, helping to pull on a rope to calve a cow, mother would let us sleep late and miss school.

After primary school we went to secondary school in Coniston. It was quite a journey by minibus, about forty minutes each way. We got to know our fellow passengers really well, and it was on this bus that I first met Miles.

Miles was two years older than me, and sat quietly on the bus gazing out of the window the whole time. He seemed unaware of the many human dramas that were played out on the bus as we changed from schoolgirls with ribbons in our plaits to young women with hairspray and lipstick.

Miles seemed totally absorbed in the landscape through the bus windows, taking a keen interest in who was going where on what tractor, which sheep were with which tup in the fields, who had got their hay baled and whose grass was down on the ground getting wet. The bus journey was a perfect opportunity for him to observe farming in the Crake Valley, and consider people's choices and decision making. He was lost in his own little agricultural world.

In my final year at school I met up with Miles again. He had left school and had been working on his parents' farm for 18 months. His hands were black with grease, his fingers thick from getting in the way of hammers and trapped in gates. Miles was on the committee of the Field Day that we were organising at Young Farmers and so was I. We had to work together to organise the event, and he suggested meeting up one night after school to discuss the details.

We arranged that he would come over to my farm, and I was watching out of my bedroom window for him coming across the fields. At the given time a small red tractor came into view, and I could see Miles approaching as the tractor bobbed up and down as it hit stones in the tyre tracks. I felt a strange sense of anticipation.

My father was in the yard, and he opened the gate for Miles to drive in. I could see them discussing the tractor, and before long the bonnet was off and Dad went across the yard to fetch a tool. The dogs in the yard ignored Miles, always a good sign. They settled down to sleep in the late afternoon sun. It was as if Miles belonged here already.

After the tractor had been tinkered with, Miles came into the house, hitting his head on a beam in the kitchen, and Mother made us a pot of tea and left us alone in the kitchen to discuss our plans. Really, that was that. Miles became part of the family. He helped when he could with haymaking and calving, and when the question arose of what I was going to do when I left school Miles asked me to marry him.

I didn't want to leave home just yet, so we decided that I would continue to live at home and help out on the farm until the following spring when we would be married. I would be seventeen and Miles nineteen.

We would live with his parents at Nibthwaite, where Miles was well on the way to taking over the farm already. His parents were a lot older than mine, and ready to pass the farm onto the next generation. They would continue to live with us in the farmhouse, but to all intents and purposes Miles and I would be running the farm.

It never crossed my mind to look at other career options, or part time jobs, as there was always more than enough work for everyone to do on a hill farm.

On Easter Saturday 1988 I left home and became Miles's wife. Miles and I began married life together in his childhood bedroom in his farmhouse. We didn't have a honeymoon as it was lambing time and everyone was busy. Going somewhere else during lambing was unthinkable. Neither of us lacked passion or enthusiasm for farming, or for each other. We were the perfect match, and looked forward to building a family together.

7

It may seem strange to you that we started married life living with Miles' parents, but to us it was the most natural thing in the world. A farmer needs to be near his animals, and so we needed to live on Miles' farm to check on cows calving, and sheep lambing during the night. It was unthinkable that my in-laws would move elsewhere as they had lived on that farm since shortly after they were married just after the end of the Second World War.

At first Miles' mother Eileen continued to do most of the jobs around the house and I helped her out. There was always something baking in the range. Miles and his father George came into the house regularly to eat, whatever the season.

There was a quick breakfast to prepare around 6am before the men started work, then second breakfast of eggs and bacon at around ten o'clock. Dinner was usually at around one o clock, and was always a hot meal with a pudding, and tea was at five in the afternoon. Before bed there was another round of tea and toast.

Between meals Eileen and I had jobs to do, some inside and some outside. In the morning I had to feed the hens, ducks and geese and let them out. Depending on the time of year there might also be lambs and calves to feed. The dogs needed exercising if they were not being worked that day, or had puppies, and they needed feeding and their

kennels cleaning out. After dinner we did our housework, washing, ironing and cleaning. In the evenings we sewed or knitted.

During my second winter there Eileen had a fall in the yard on some ice and cracked three of her ribs. After that, by unspoken agreement, I took over the outdoor jobs and Eileen still helped in the house in the afternoons. I noticed that George would not always go out on his rounds with Miles, sometimes he would stay in his workshop making a stick from a tup's horn or mending a piece of machinery, if it was particularly cold or wet. Both of my in-laws eased gently into some sort of retirement over the next five years, although retirement was never spoken of or mentioned.

The tenancy of the farm was officially signed over to Miles on his 25th birthday. This significant handing over from one generation to the next was celebrated in our local pub with a glass of beer, and the thought of the next generation began to weigh heavily on my mind.

When I first moved in with Eileen and George, Eileen would regularly speak of knitting baby clothes soon, and making the box room into a nursery. Over the last couple of years the talk about babies had stopped. My brother had married three years after me and already had one child and another on the way. At least the future of my parents' family farm seemed to be in safe hands. My father had passed his tenancy onto Robert, and Robert and his wife June now ran the farm and it was their family home, and their child's future.

I had tentatively broached the subject of why I wasn't pregnant with Miles. Both of us had been involved in animal husbandry for our entire lives, and it grieved me to think that I might not be fit for purpose like a geld yow. A geld yow might be given a second chance by a generous farmer, but generally if she had not produced a lamb from her second mating she was not worth keeping in the flock and would be sold as mutton. I felt like a useless lump of mutton, unfit for breeding, not worth my place in the flock.

Whenever I broached the subject with Miles he replied that 'Nature would take its course' and that was the end of the conversation. We were hill farmers, and I wondered if had we been dairy farmers used to artificial insemination if the conversation would have been different. Our lives were governed by the seasons, nature and its fertility. Fertility was the key to our success as farmers, but somehow Miles had chosen a geld yow for a wife.

The whole point of a hill farm it seemed to me, the whole reason for its existence, was to improve the flock and pass on the flock in a better condition to the next generation. What if there was no next generation? The landowner could take possession of the farm and pass the tenancy over to someone else, or sell the land, flock and house separately and the farm would cease to exist. Another piece of the jigsaw of Lake District farming would be lost.

In my dreams, Miles and I stood frequently on the top of Bethecar Moor with suitcases in our hands. The continuation of the family farm depended upon me producing an heir, and I didn't seem to be able to do it. It grieved me every day not to have a child. There was no child to help feeding the lambs, no child to receive an Easter egg from the vicar in church on Easter morning, no child to visit Father Christmas in his grotto. Without a child what on Earth was all this work for? We had a three generation tenancy from the conservation organisation, but no third generation to take it on.

8

At the time of my 40th birthday it seemed to me that there was very little to celebrate. Turning 40 involved for me an acceptance that I was never going to be a mother, and that we would have to leave the farm once we passed retirement age, as there were no children to take it on.

My father-in-law George had recently died after a short illness. Burying him felt like burying part of the farm, and it was difficult to work for a future that did not seem to exist. Miles carried on cheerfully working each day, stoically outside with his flocks in all weathers. He had been born to the work, so there was really no option. Ideally he said that he wished to die on the fell like an old red deer stag, but life was not always that straight forward. Eileen was now not in good health, and largely housebound, and relied upon me for news and conversation. I brought flowers from the valley into her bedroom to cheer us both up, but we knew that our hearts were both heavy as I did not have a child.

Eileen suggested that I try a new project to take my mind off the future, and together we talked about a variety of farming diversification ideas. The one that seemed to be the easiest to set up and the most straight forward to run, whilst remaining at home caring for Eileen and cooking for Miles, was to set up a holiday let.

We have on our land a very old farmhouse at

Parkamoor. It's watertight, and basic, and is still furnished from the last tenant who died up there in a snowstorm. We decided that I would do an 'off grid' holiday let.

I tried the holiday let idea for two years, but to be honest our holidaymakers were just not equipped to cope with off grid living. They seemed unable to look after themselves and would be forever turning up at the farmhouse saying that they were unable to keep the fire going, light the paraffin lamps or pump water. Several asked for refunds saying that it was an unsuitable location for a holiday let, some left very rude comments online, so the holiday let idea was put to rest, and it was just as well as shortly after that Eileen took a turn for the worse following a stroke. She needed me more than ever now.

From time to time neighbours would come in to sit with Eileen and chat, and it was from one of these neighbours that I heard that Sawrey Moor was staying up in our farmhouse. I had first met Sawrey when she was still in her pram, when I still had hopes of being a mother myself, and had taken an interest in the tiny scrap of a girl wailing in a knitted bonnet.

Sawrey was now a fully grown woman, a woman who was apparently living in our holiday let. I asked Miles to go and tell her to move out, but he said that every time he went up to the house she was never there. I knew when to keep quiet, and I knew that Miles would sort out the situation in his own way. I trusted him, it was his farm and he could sort out a squatting young woman, a situation I thought that I was best keeping well out of.

Autumn is a very busy time on a hill farm before the winter sets in. We call it the 'backend' and there are lots of jobs to do. The hay or silage must be made when the weather is right at the end of summer, and the bales stacked and stored ready for the bad weather coming. The lambs must be weaned, or 'spained' as we say – they are separated from their mothers and sorted through. The best gimmer lambs will be kept as replacements for the flock. The other gimmer lambs will be sold at a breeding sheep sale such as the Lakeland Fair at Kendal auction. The lambs need to be looking their best for these sales and I would be drafted in to help.

The castrated boy lambs, the wethers, will also need to be sold during September or October before the grass stops growing for the winter and we have nothing to fatten them on. These are called 'store lambs' and are sold through the auction to a farm with better land to fatten and sell for meat. Most upland farms sell their wether lambs 'store' as they do not have the resources to fatten them themselves.

The gimmer lambs that are to be kept as replacements do not immediately re-join their mothers in the flock. To help them grow on they are wintered away on a farm with better grass. Ours go to a farm near Barrow in Furness and we have to take them there and then visit them periodically to see if they are alright.

The yows themselves also have to be sorted through. Any with bad udders or who have lost their teeth are considered unsuitable for breeding again and are sold 'cast' at the auction and will go to make mutton.

Any yows that have done four years on the fell and look unlikely to be able to do a fifth comfortably, but have reasonable teeth and good udders will be sold as 'drafts' at a draft ewe sale. Draft yows are generally those that are correct in udder and mouth and have reared lambs successfully but would be better moving to another farm with a kinder climate for the rest of their breeding lives, usually another two or three years. These are highly sought after by farmers who will buy them to cross with a Texel or a Suffolk, a meat producing tup, in order to breed fat lambs for supermarkets. Buyers know that if a sheep has survived for four years on the fell she is pretty hardy and will do well on a lowland farm or a valley bottom farm.

So with one thing and another, there is always a lot going on in September and October, and my husband mentioned that he was going up to Parkamoor to take Sawrey some hay for her Herdwick sheep I was mildly irritated that he was doing this for her, who paid no rent, at this busy time, but saw no reason to worry.

As it started to snow I was cooking tea for Eileen and Miles. It came on really quickly, as it sometimes does, and I kept listening for the sound of the quad bike. When it got to 9pm and he still wasn't home I was really worried, but Eileen rightly said that if there were any problems one of his dogs would have come back to tell us.

After supper I fell asleep reading to Eileen lying on the bed next to her. The next thing I knew it was morning and a strange white light was filling the room from the snow outside, and I could hear the sound of the quad bike moving about in the yard and the dogs barking wanting their breakfast. I assumed that Miles had come back after I fell asleep. We were so busy that day feeding animals in the snow that I never actually asked him what time he got back from Parkamoor and what had happened there.

9

Eileen loved her visits from neighbours, telling her the local news, as she lay bedridden looking out of the window across Bethecar Moor. One came with the surprising news that Sawrey Moor was now visibly pregnant, still riding her pony and living at Parkamoor. Eileen passed on the news to me, and I stood with my back to the Rayburn looking out across the farmyard waiting for Miles to tell him.

We would have to tell Sawrey that she had to go home. Honestly, I could not be held responsible for a pregnant woman living in one of our houses that was completely inaccessible without running water or an inside toilet. She might need medical help and it might be difficult to get a doctor to her – Miles would have to tell her to go.

Miles came into the kitchen looking for a cup of tea. He took off his wellies and his waterproof trousers and left them standing in the doormat.

'You'll never guess what your Mum has just told me.'

'What?' He seemed disinterested, looking for his tea.

'That Sawrey Moor is pregnant. Bold as brass, riding her pony with a big bump. You'll have to do something about it Miles, tell her to go, because I can't have her living up there like that.'

'What?' he said again, all the colour had drained from his face. He went over to the waterproof trousers and wellies, started to put them on, then had a change of heart and sat down.

'Sawrey Moor is pregnant. Do you not understand? I can't have her living up there miles away from a road when she might need a doctor. You'll have to tell her to go.'

'I can't,' he said.

'Why not?' I asked. 'This changes everything. I can't have her up there, pregnant, who knows what might happen to her.'

There was a silence for a while as I waited for a response. He didn't say anything so I turned to slide the kettle across from the simmering plate on the Rayburn to the boiling plate to make the tea.

'I can't say anything to her,' he said, 'because the baby will be mine.'

With that sentence my whole life changed. I dropped the kettle onto the hotplate, ran over to the sink and was physically sick. I didn't doubt for a second that it was true,

as Miles would not have said that if it wasn't. I slumped to the floor, with a tea towel held over my face. He came over to touch me on my shoulder

'Do not touch me!' I shouted. 'You have no right to ever touch me again.'

In the days that followed I walked around doing my jobs in a trance. I could not leave because Eileen was completely dependent on me. I delivered food to her with a smile, and read to her, but inside I was broken.

I felt as if someone had sliced into my chest, cut out my heart, thrown it on the floor and stamped on it. I am sorry to be so visual, but my pain felt physical as well as emotional. I was walking around, but I felt as if Miles had killed me.

I extracted every detail about the evening from him. I made him go over and over the details trying to work out who was to blame, trying to work out what to do. The best part of twenty years of my life had been spent in trying to provide the next generation of farmers for this farm, and now Sawrey Moor had done in one night what I couldn't do in twenty years. I really was a geld yow, barren, not a breeder, and should have been sold to slaughter years ago. I would have saved myself a whole lot of heartache.

10

I heard that the baby had been born from a neighbour. There was no discussion that I was involved in about who was the father, although I knew that they must have been discussing it in the valley. I did not discuss the baby with Miles; I didn't want to know if he had been to visit.

I had gone over everything a thousand times in my head, and I knew what I had to do. I was preparing for it, sorting out clothes for the charity shop, clearing out cupboards, cleaning the house. I was virtually housebound myself, as Eileen was now so poorly. She had refused the move to a hospice, and was now eeking out her final days at home.

I don't know if Miles had told her about the baby, but she never discussed it with me. We all lived in the same house, going about our daily business, with no real conversation. After I had grilled Miles for all the details, and decided what to do, I began to distance myself from him emotionally, steeling myself for the future.

Eileen died on a stormy day, taking her last breath beside me as the rain hammered on the window. She turned her face to look at the moor, and died. I called the doctor, but refused to have her taken to a chapel of rest. She wanted to be at home. I washed her body and laid her in the bed waiting for the coffin I had ordered to come.

When the coffin arrived Miles put it in the hardly ever used living room, and we carried her body through. She was dressed in her favourite tweed outfit. We padded the coffin with a soft woollen blanket and laid her in. Together we said goodbye to her, and Miles fastened the lid on the coffin.

There was still three days until the funeral so I brought some foliage in from outside and put vases around the room. On the day of the funeral Miles strapped the coffin to a bale trailer with ratchet straps, and I stuck some foliage under the straps to decorate the trailer. My brother and his wife came over from their farm, Miles drove the tractor pulling the trailer and I sat in the rear seat of my brother's Land Rover.

It was a very still, cold, sunny day. We followed the tractor down the hill watching anxiously for any movement of the ratchet straps. Once at the church, Miles parked right outside and jumped down to loosen the straps. We stood and waited. Miles, my brother and two local farmers carried the coffin into the church, and I walked behind with my arm slipped through my sister-in-law's. As we entered the church she whispered to me

'Are you really going today?'

'Yes,' I said.

Now was really not the time to be discussing future plans. We had to focus on Eileen.

Afterwards in the pub, once most of the sandwiches had been eaten, and the men looked settled for the evening at the bar, I slipped quietly away and began to walk up the hill back to the farmhouse. Everything looked beautiful today, as things often do when you know you must leave them. I was glad that I had worn sensible shoes as I walked along the field edges to the farm. I could not look at the sheep in the fields, knowing I had worked alongside so many generations of the flock I now had to leave. I could tell you each sheep's character and breeding, but I could not think of that now. One or two that had been bottle-fed as lambs insisted on walking along next to me, nibbling my skirt.

'Nothing for you today girls,' I said. Nothing much for me either.

I loaded my packed bags into the Land Rover and said goodbye to the dogs. I took a last look around the kitchen, and put my letter next to the teapot on the table. I had failed to produce the next generation, but nature had found a way to provide it. Nature always finds a way. Now I had to leave in order to make way for that next generation.

I was being drafted, sold on to an easier climate. Like the draft yows, I hoped that I would thrive in my new surroundings. All that I left behind was a dignified silence.

We had two Land Rovers, one had been Miles' father's, and despite being a lot older it was considerably cleaner. It was this one that I took. I turned the key and the petrol light came on.

'Bloody Hell,' I shouted. 'Bloody Hell. Why does the petrol light always come on when I want to go somewhere and it is fifteen miles to the bloody petrol station?'

I put both hands on the wheel and years of tears and frustration came flooding out. As a farmer I was used to holding my emotions in and getting on with things when a favourite yow died or a lamb drowned in a stream. I could hold it in no longer. Driving through the tears I began the bumpy route across the fields.

I had confided to a friend in Rusland about Sawrey, the

baby and my dilemma about what to do next. She had mentioned to me a few weeks ago that she was stopping letting her farm cottage as a holiday let and was looking for a tenant. It had become a bit of a chore for her doing the weekly changeovers, and sometimes holiday makers with their barbeques, music and really annoying habit of relaxing and doing nothing was incompatible with life on a farm. If you were filthy, exhausted and grumpy after being up all night lambing, the last thing you wanted to see was a couple lounging around in your garden relaxing.

She was concerned about what sort of tenant she would get, and I asked if I could rent the cottage. The rent was to be four hundred and fifty pounds a month, so all I had to do now was find a means of financially supporting myself. I had never had to do this in my life before.

I considered what my skills were. I could cook and clean, neither of which I particularly enjoyed, and I could do farm paperwork. I had done all of our farm administration for about ten years now. I was slightly hesitant about applying for a job I saw advertised at Kendal Auction, in the office, and I fully stressed all of my weaknesses at the interview, but I was delighted and terrified in equal parts when I was offered the job. I was to start my job the following week. Part time hours, uniform provided, ability to communicate with farmers a must. Such a lot of change was happening in my life.

I drew up outside the cottage in the Land Rover. My friends would still be at the pub enjoying the wake. The house was a typical Cumbrian longhouse, where all the rooms led into each other without internal corridors, and the farm cottage was attached at the end.

There it was, my own front door. I rummaged in my bag for the key and let myself in. I had a small sitting room and kitchen downstairs, and a bedroom and a bathroom upstairs. As the cottage had been a holiday cottage everything that I needed was there, the bed was even made up ready for me. I hung my coat on the coat pegs and went through to the kitchen to make tea. It was very strange to be in an empty house, no family, no farm cats pawing at the window.

My new life had started. I was an independent working woman with her own home. I could never have predicted that this would happen. I had never even considered that I might leave the farm until I died. There was a whole new world without farm work waiting for me to enjoy out there. For now I just sipped my tea and looked out of the window. A new view, I would never have predicted that.

11

Miles:

'I never meant for any of this to happen. It was just one day when Nature took over. I've watched animal behaviour all my life, and Nature will always find a way to reproduce.

'I'm sorry for the hurt I've caused. I'm so sorry that I have hurt my wife, but I honestly didn't set out to hurt her. It just happened. I've learnt the hard way that one moment of madness can have a big impact on your life.

'I'm not sorry about the baby though. How could I be sorry about that? I don't know what the future will hold, but now I have a child it is a different story. I'm not sure how it will end.'

12

Sawrey:

'I won't say it was easy settling in to being a mother, and I won't say it was difficult either. Suddenly living up at Parkamoor did seem to be wildly impractical. I had no washing machine for a start, so I decided to buy some disposable nappies, but I had no bin either. There were lots of practicalities to sort out, but my day-to-day existence remained the same.

'Bee and I still checked the sheep. If we needed to go into the village we had Apple, and I would walk alongside her with Bethecar strapped on to my front in a sling. She was so tiny that at the moment this was practical, and Apple and I were such a team that there was little chance of her knocking into the sleeping beauty.

'I wondered if Miles had heard the news about the baby, and how he would react. My mother had told me that his mother had died recently, and I guessed that sorting out the funeral would take up most of his time at the moment.

'I didn't have to wonder long as I heard the quad bike coming along towards the house. My heart started beating very fast and I took several deep breaths. I stood in the porch holding Bethecar sleeping against my neck, and suddenly I felt as if I was falling through time. In my mind's eye I could see the generations of women who had nursed babies in their arms waiting for fathers coming across the fell. Something was changing in my life; I could feel it like the wind.

Miles turned off his engine, and without saying a word walked towards me holding out his hands.

'Come inside,' I said. 'There's a cold wind out here.'

We sat side by side on the settle next to the fire, the flames making our cheeks burn after being outside. I handed Bethecar to him, wrapped in her cosy blanket, and he took her and held her close.

I was not in the habit of seeing men cry, but now the tears were rolling down Miles' cheeks as he sat in silence looking down into the face of his child for the first time.

'I'm sorry Sawrey,' he said, 'I thought this day would never happen, I thought I would never be a father.'

He held her in silence, and then passed her back to me while he paced about on the stone flags gathering his thoughts.

'I don't know if you know, Sawrey, but my wife has left me. She has left me so that I can take care of the baby, the next generation, and I would like to do that Sawrey. I would like to look after you and the baby, and I would like you to move into my farmhouse so that I can look after you both properly. I hope in time you will come to care for me too, but the most important thing is for the baby to be in a loving home with her parents. What do you say?'

I was absolutely dumbfounded. I had never heard him utter more than a brief sentence before, and now he had made this declaration of desire to care for Bethecar and me. I had moved here with the desire to be an independent fell farmer, with my own hefted flock. I had not anticipated starting a family, and definitely had no desire to become a

farmer's wife cooking and cleaning for a husband.

'I don't know what to say Miles,' I said.

'But you're not saying no?'

'No, I'm not saying no. Give me some time to think about what is best for me and Bethecar will you?'

'Of course' he said, 'I'll be back at teatime tomorrow and you can tell me what you have decided. And Sawrey, I hope you don't mind me saying that I have never seen a bonnier baby, or a bonnier mother.'

With that he put on his boots and left. It was an unusual situation for parents to be in, but I had to work out what to do for the best, by tomorrow teatime.

13

The following day Miles came back up on his quad bike, towing a small trailer. 'He's assuming I'm going back with him then' I thought. I was feeding Bethecar in a chair by the window, and waved to him to come in.

'Have you decided?' he asked. Quietly, with Bethecar feeding at my breast, I explained my thoughts to him. I agreed that it would be best for Bethecar if we were both together, and I thought it was worth seeing if we could get along together. I had to be free to go back to Parkamoor at any time that I wanted. This was important to me; the knowledge that I had somewhere else to go and was not obliged to live with him if I didn't want to. I was to run my own farming business, and my own flock of sheep. You should have seen his face when I announced this. All decisions relating to the flock were mine alone, all the work at Parkamoor farm with my sheep was to be done by myself with Bethecar. Lastly, Bee was to be allowed to live inside the house, and Apple in one of his stables or in a paddock close to his house.

'It's a yes then, is it?' he said with a smile on his face. 'Let's get the trailer loaded up while it's dry.'

He helped me load everything I needed for Bethecar and myself into the trailer, and set off for home. Bee, Apple, Bethecar and myself walked behind.

'It's not goodbye to Parkamoor,' I said to Bee, 'we will

have our tea breaks here every day.' I wasn't sad to be leaving; as the situation was so unexpected I think I was still in shock at what was happening.

There was rather a disturbance when we arrived at Miles' house, with the dogs barking at Bee and the unfamiliar sound of hooves in the farmyard unsettling everything. 'They'll soon settle down,' I thought, and went into the farm kitchen.

I had never actually been in the farmhouse before. Of course there was a Rayburn keeping the kitchen cosy, and Miles slid the kettle across. I sat down on one of the chairs unsure as to what to do next. Miles nervously explained that he had cleaned out his parents' bedroom, and I could sleep in that room with Bethecar. He had his own bedroom just across the landing. He wanted me to know he was taking nothing for granted.

He made me some tea and toast, and asked if he could look after Bethecar while I had a bath. A bath with running water! Such a luxury after living for so long without water on tap. He showed me up to the bathroom and made sure I had towels, soap and shampoo. I handed him Bethecar and settled down in the bath. I was still unsure what Miles expected the domestic situation to be, I had taken the time to explain to him in no uncertain terms that I did not like cooking and cleaning.

After my bath I settled Bethecar in the tiny basket that she slept in next to the bed. We had brought the old skep basket back from Parkamoor. I had found it in the storeroom, I don't know what it had been used for previously, it was a sort of swill basket made from oak..

'I'm going to bed now,' I said, and slipped between the

sheets. There was no conversation, but Miles lay down next to me on top of the bedclothes and held my hand while I went to sleep. I woke during the night to feed Bethecar, and looked across at Miles lying fully clothed on top of the bed-covers, fast asleep. As it got light and the sunlight poured into the room I was sat up in bed feeding Bethecar. Miles woke and looked up at us. He kissed us both on the head and went off to start his work for the day. A new day, a new beginning, a new life.

14

It wasn't a difficult transition to make settling into Miles' routine. I still travelled up to Parkamoor each day, on a quad now not on foot, and ran my own flock. I had my own sheep, and my independence in my work, but I had somebody to share Bethecar's little milestones with. When she smiled for the first time we both saw it happen, when she sat up on her own we both gasped, and when she took her first tentative steps it was across the farmyard towards Bee.

I managed the flock on my own and took my own sheep to the auction, and bought and chose my own Herdwick tup from Broughton Auction. I was very surprised to learn from a fellow shepherd that Miles' wife was working at Kendal Auction, although when I later when into the office she moved away from the counter and somebody else served me.

For the first time I thought about how difficult it must be for her to see Bethecar and me at the auction. Miles had just said that they had drifted apart and she had left, and I had accepted this explanation at face value. He said that they had been unable to have children and that was what had ended the marriage. Seeing Bethecar, hearing her shouting across the auction office, must have been difficult for her.

I had no other option than to go into the office. I couldn't just avoid paying, but I realised now how difficult it was for her to see myself and Bethecar, and my once again swollen pregnant belly coming into the office.

I had been living with Miles for a year now. Our desire to be a family, and our desire for each other, had meant that he had never actually gone back into his own bedroom. We quietly became a couple, without discussions of the future, because our future was already tied up in Bethecar. I was delighted when I discovered that I was expecting another baby. I felt that somehow it would complete our family. Miles was also delighted, although he knew better than to make a fuss of me. We both got on with our own work, focussing on the tasks that had to be done that day. There was little time for relaxation and discussion, but we were both content, busy with our flocks and expanding our family.

The end of the summer is such a busy time on a hill farm. Decisions have to be made about which young sheep will be kept as flock replacements, and the remaining young sheep sold before the grass stops growing and winter starts. I made all these decisions with the Parkamoor flock. I often put Bethecar down for a nap in the old farmhouse while I went outside to my roughly constructed sheep pens to worm the sheep, or check their feet.

I had bought one tup at Broughton Herdwick tup sale, but my flock had expanded to over 50 yows now so I thought it would be a good idea to get another tup, if only to 'chase up' those who did not get pregnant by the first tup. The Lakeland Hill Fair is an 'end of season' sheep sale held each year at Kendal Auction and the catalogue arrived in the post. There were over 2,000 breeding sheep for sale in one ring, and a selection of tups for sale I the other.

I noticed that one of my neighbours was selling an old

Herdwick tup that I had admired and decided to go and hopefully buy it. It is a good idea to buy a tup from a neighbouring farm as it shows the tup can survive the local environment. We have a lot of ticks on Bethecar Moor, and a tup that was not used to that might die.

So it was that Bethecar and I set off to the Lakeland Hill Fair on the third Saturday in October in the Land Rover with the small sheep trailer to bring the tup home in. We arrived just before the sale was due to start, and there was hardly anywhere to park. Land Rovers, pickups and trailers had been squeezed into every available space at the auction and it was difficult to find somewhere to park. I ended up parking on the ramp down from the roundabout into the auction complex, as there was nowhere else to park. Inside the auction the familiar smell of sheep and sawdust filled the air, sawdust sticking to the soles of my boots.

We found a seat to sit looking down on the ring, and before long I could see my neighbour in the pens behind the ring waiting to come in with his old tup. Eighty pounds bought the tup, and I stood up ready to go. I had Bethecar on one hip as I slid past everyone in the other seats to get out, and walked down to the auction office to pay and get my movement forms.

I began to feel decidedly unwell and thought I must not have eaten enough breakfast. Once inside the auction office I put Bethecar down on one of the chairs and joined the queue. It's difficult to say what happened next, but I was filled with pain from top to bottom, and a feeling that I needed to go to the toilet. I crouched down, and the next thing that I knew I was sat on the floor, with fluid leaking between my legs. I was thinking, 'oh no, don't let Bethecar see me like this'.

There was a general commotion as some of the men in the queue shouted that an ambulance needed to be called for, and Ruth had rushed from behind the counter and was sat on the floor next to me. I looked across at her and our eyes connected. Two women fully absorbed in the situation, despite our history.

'I'm losing the baby, Ruth,' I said.

'The ambulance is coming,' she said 'and I'm here with you. I'll ring Miles.'

I sat on the floor leaning against a chair, with Bethecar swinging her legs behind me. I could hear Ruth on the phone to Miles, telling him what was happening, and to my brother telling him in no uncertain terms that he had to get to the auction as soon as possible to pick up Bethecar, my Land Rover and the tup. I could imagine the cursing that would be coming down the line at her.

The conversation ended with her saying, 'We'll leave Bethecar here in the auction office with the office staff, her car seat is in the Land Rover, for goodness sake just get here quickly.'

Before long the paramedics arrived. Questions were asked and answered, and they decided to take me to Lancaster Hospital. To my utter amazement I heard Ruth saying, 'I'm the baby's aunty, I will be coming in the ambulance.'

The drive to Lancaster was not an easy one. I held on tight to Ruth's hand, making a series of little crescent shapes in her hand with my fingernails. By the time we got to Lancaster I had delivered a tiny baby boy who did not take a breath. The boy was wrapped in a blanket, and given to me to hold, and we travelled along at high speed in a world of our own where time stood still, falling and turning around me as the paramedic took various readings and asked various questions which were answered by Ruth. I was now lost in the moment, with silent tears running down my face, unable to speak.

Shortly after we arrived at the hospital Miles appeared and there was then a whispered conversation between him and Ruth. I could see by the way Ruth talked to him that she was still in pain after their separation. Ruth then left, and Miles came and sat in the chair by the bed, his big dirty boots and hands strangely out of place in the sterilised environment. I looked at my baby boy's tiny, perfect hands, and at Miles' finger ends that were big and splayed, bashed by a thousand hammers and caught in a thousand gates. My boy would never hold a hammer or close a gate. I closed my eyes and wished I could sleep, but when I did all I could see was Ruth's face. How on earth had I imagined that it was acceptable to come between her and her husband?

15

Ruth:

I had settled well into my new life away from the farm. I had my own little house and nobody but myself to please. I could lie in bed in the morning if I wanted to, or get up early and wander outside and down the lane to get a breath of fresh air. I had been used to the outdoor life, and it was quite difficult to adjust to not needing to be outside for some of the day.

To give me an excuse to go out walking, and for company, I decided to get myself a little dog. A friend who bred terriers offered me a little bitch that had been unsuitable for breeding. She had had one litter of puppies but had to have a caesarean, so wasn't to be bred from again.

She arrived about a month after I had rented the house, badly needing a haircut as she could hardly see out from behind the hair around her eyes. She had lived in a barn all her life; the same as thousands of breeding dogs, so I had to teach her what was and was not acceptable in a house. Thankfully she proved easy to train, and would make it her mission to please me. She already had a name, Erika, goodness knows why, but she wasn't really used to people using it.

Erika and I became a team. We walked together through the lanes of Rusland before work, and then she came to the auction with me and waited in the Land Rover while I worked in the office. I found the work enjoyable, not difficult, and I suppose the same could be said generally of my life. It was enjoyable, not difficult. Gone was the requirement to work everyday on the farm, and to prepare food for people, my life was my own. At the weekend when there wasn't an auction Erika and I went walking around Coniston or Tarn Hows. All in all, things had worked out pretty well.

I saw Sawrey regularly at the auction, and when I realised that she was pregnant again I thought 'well, that has to be good news for the farm'. I hadn't seen Miles for ages. He must have taken his sheep to Broughton Auction, no doubt Sawrey had told him I was working at Kendal.

The Lakeland Hill Fair was always one of the busiest days of the year at the auction. I had experienced it many times both as a buyer and seller, and now as an employee. We had to create accounts for new buyers before the sale, and take payment and give out receipts and movement forms after the sale. Everybody worked on Lakeland Hill Fair day, so there were eight of us in the office.

I was busy with a customer when Sawrey and Bethecar came in, but I heard Bethecar singing and glanced up. I looked again, and Sawrey was collapsed on the floor. Straight away I could see it was a serious situation. All hostility between Sawrey and myself was irrelevant, at this point she was a woman who needed help.

I waited with Sawrey while the ambulance came, and organised someone to look after Bethecar and collect Sawrey's sheep. I just had to leave Erika in my Land Rover in the car park; there was nothing else for it.

Sawrey was so brave in the ambulance, and my heart went out to her. Nobody should have to suffer losing a child alone, and my instinct was to stay by her side. I wondered for the first time what it was like for Sawrey to give birth to Bethecar at Parkamoor. Was anybody with her? Surely she couldn't have given birth up there alone without even any running water?

Of course, she had got herself into that situation, but I had never seen anyone give birth before and it had never occurred to me that it must have been a very frightening time for her. There were lots of questions to answer for the paramedics, and Sawrey was in a lot of pain. I phoned Miles again from the ambulance to check he was on his way, and told him in no uncertain terms that he was expected as soon as possible at the hospital. I gathered that he hadn't yet left the farm, as I don't think he would have answered the phone if he were driving. It hadn't occurred to me at this time that Sawrey had the Land Rover at the auction, and I had his parents' old Land Rover, and Miles was at that moment trying to borrow a car to get to the hospital.

When Sawrey delivered the baby I thought that my heart would actually break as he was wrapped in a blanket and passed to her to hold. I thought that I had been through a traumatic time in the last few months, but it all paled into insignificance now. Sawrey kissed her son and sat, pale faced and silent, waiting for the journey to end.

I had the utmost admiration for her. I hated to see her in pain. I felt that it should have happened to me and not her, to spare her the pain. I had seen a lot of pain and both life and death on the farm, but this was the worst thing I had experienced, and I would make sure that Miles knew that when he arrived at the hospital. Sawrey needed support now, not from me but from my husband.

16

Miles:

Living on a farm we are well used to life and death. Sometimes it is just too much of a struggle for an animal, like our poor little lad who never got to take a breath.

It's really knocked Sawrey for six. I didn't think it would upset her so much. I think she's upset because she wanted a boy to take over from me, but honestly I'm very happy to have Bethecar.

It was all very awkward with Ruth being there when it happened, but by all accounts Ruth looked after her and there wasn't a cross word. What a woman, I was very lucky to have her in my life for so long.

Life goes on, spring must follow winter, and I hope Sawrey feels better soon.

17

Ruth:

I left it about a month after the miscarriage before going to see Sawrey and Miles, but in my mind the events of that day had brought things to a head. We could not carry on as before.

I drove into the farmyard. It felt as if I had been away for years. Nothing stirred. No dogs barked. They knew the sound of the old Land Rover. I would not go and see the dogs; tonight would be difficult enough as it was. It was one of the dark days of December, when daylight is in short supply and nothing ever thaws out properly during the day. The farmyard was covered in sheet ice, but two roses bloomed around the door. Christmas roses, bringing cheer to midwinter days like Christmas itself.

I knocked on the window in the kitchen door and Sawrey came to the door and let me in. She looked exhausted, as we all do during December, and after we had said hello I looked around the kitchen. Miles was sat in his chair by the Rayburn, drinking a cup of tea. He did not get up. The kitchen looked very different from when I was living there. There were toys and cooking implements all over the kitchen floor where Bethecar had been playing. Nothing seemed to be in the right place, and there were airers with washing on parked in odd places all over the kitchen. I focussed my mind on why I had come here.

'I just wanted a word with Miles, Sawrey, if that's all right?' I asked.

'Of course,' she replied, 'I was going to take Bethecar upstairs now anyway. Say Night Night Bethecar' and off they went to settle Bethecar into bed.

I went and sat in the chair on the other side of the Rayburn to Miles, where I had spent so many evenings sitting in quiet companionship with him.

'Miles, I think you and I should get divorced so that you can marry Sawrey. Sawrey needs your support, especially after what she has just been through, and we need to get this sorted out before Bethecar starts school so that nobody says anything about her Daddy being married to somebody else. It's only fair on Bethecar.'

'Yes, you're right,' he said. 'We need to do what's best for Bethecar.'

'As you know I've been doing the farm accounts for years, and I've worked out a way that we can split our assets. We don't own any land, so that's not a problem to sort out. I will just get my name taken off the tenancy agreement. We have two Land Rovers and I've taken one, so that's also fine.'

He looked quite relieved, perhaps thinking he was about to get off lightly.

'As for our other assets; tractors, trailers, livestock, I've totted up the amounts as best as I can here…' I opened up an exercise book with all of our assets neatly ascribed a value.

'Of course I know that you need all the farm equipment, and I couldn't expect you to sell your hefted flock of sheep as it would be such a huge job to start again. So, what I think is the way forward is for you to sell your cows. This would give you enough to pay me fifty per cent of our assets. I know your income would be dramatically reduced, but I really think that this is the easiest way, Miles.'

Miles was now on his feet, 'Sawrey, Sawrey, come down here now.' Sawrey came rushing down the stairs looking concerned.

'Ruth says I have to sell the cows so that we can get divorced.'

'Well Miles,' she said with a smile 'something had to give.' I saw a glint of happiness, a hope of a future with Miles in her eye, and I know that between the three of us we would work it out.

18

Old Malcolm Postlethwaite had come into the auction office one day to pay a bill and waited until I was available to help him. He had been my parents' neighbour for as long as I could remember, and had no children to take on his farm, so was keeping on as well as he could by himself into his eighties. I knew that my brother regularly went over to help him out with things.

'Hello Ruth,' he said, 'I've mentioned to your brother that I'm going to be selling my two meadows, you know the ones that border onto his farm. I hoped he would buy them, but he says he hasn't any money.'

'Really Malcolm?' I asked 'the meadows at Nibthwaite with the barn in?'

'Yes, I'd have liked your family to have them, seeing as we've known each other so long.'

'Leave it with me Malcolm,' I said. 'I'm sure we can sort something out.'

I had known Malcolm's land all my life. I had helped stacking bales into the barn on many a fine, haymaking day. In summer the fields were full of native species of flowers, as Malcolm had not fertilised his fields as much as other farmers due to trying to keep his costs down.

I knew at once what I was going to do with the money that I had got from Miles' herd dispersal sale. I didn't have enough money to buy a house, as houses in our part of the

Lake District go for a lot of money for second homes or holiday lets, but I did have enough to buy ten acres meadow.

A valuer came out from the auction to look at the fields with Malcolm and I, and a price was agreed. Within six weeks I was the owner of the meadows, a landowner that neither my father, brother nor Miles had ever been. There was a large barn in one of the fields and I had a long-term plan to convert it into a house, but for now Erika and I enjoyed the simple pleasure of walking through our own fields. My brother would farm them, and eventually they would become part of his children's farm. I could think of no better legacy to leave my family than a productive meadow.

From the meadows, on the edge of Bethecar Moor, you could see along the Crake Valley to Coniston Water. There was no view like it for me; nothing could match the familiarity and harsh beauty of that corner of Lakeland.

As Sawrey married Miles, in Lowick Church at the bottom of the valley, Erika and I stood in a meadow feeling the wind on our faces and looking down the valley at the view. I had been invited to the wedding, but wanted to stay away. That story was for Miles and Sawrey. I had no place in it. My future was here, in my valley, in my meadows but without Miles.

I had lost my husband, and my hope of ever having a child, but I felt energised by the valley itself. Owning a little part of it secured my future here. I belonged in the valley, I belonged to the valley, and my future was here, alone.

BETHECAR MOOR

Rusland

Every beech tree
Etched with graffiti
Is now etched on my memory.
Nine years of driving through the valley
On the school run twice a day.

I know the comings and goings
And the workings of this valley.
Postman at two to deliver the post,
And four to collect from the letterbox.
Milk tanker passes at 10.45.

During winter my neighbour
Will come past on alternate days
With a bale of silage on a spike.
Odd visitors to Arthur Ransome's grave
Some sat nav followers on 'shortest route'.

This is not the fastest route
To anywhere in particular.
This is a journey back in time
A time where there was real community
A time when we worked together.

RUSLAND

For nine years I have lived
Largely within the confines
Of these valleys. Between
Rusland Heights, via Bethecar Moor
To Blawith Common.

This has been my home, work
And what counts as recreation.
This valley has been my life.
It is all etched on my memory
To be recalled as and when required.

Now is the time to leave the valley
On to pastures new, farm buildings new,
Fellsides new. The future is elsewhere.
Thank you to the valleys and their farmers
May the weather be always on your side.

May you have dogs that listen,
Land Rovers that do not break down,
Haymaking summers, frosty
Backend mornings to kill off bugs,
And sheep that stay alive.

Acknowledgements

This book would not have been published without the belief of my publisher, Dawn Robertson, in both myself as a writer and in the importance of Cumbrian hill farming. I am again indebted to her for publishing my work, and allowing me to meander creatively into the world of fiction. Without Dawn's faith in me, the beginning of 'Bethecar Moor' would have remained unfinished in an old exercise book, and never would have seen the light of day.

Thank you to all the inhabitants of the Rusland and Crake Valleys. Their quiet, contented lifestyle has inspired the characters in 'Bethecar Moor'. I hope that I have done the inspirations and aspirations of the folk who live there justice. Thank you to everybody who helped me with my farming in Rusland and Blawith, especially Raymond, John and Joe who helped me on numerous occasions and whose friendship made difficult times possible. Whether it was moving large loads on tractors for me, or driving fifty miles to collect a gander to mate with my goose, their help was above and beyond what I would have expected.

Thank you to my family for supporting me with both my farming and my writing. Thank you to everyone who has bought or sold my books, to the local bookshops who have displayed my work and everyone who has encouraged me to write by commenting on Twitter and following @ruslandvalley. Our late night 'conversations' while I'm writing and everyone else in the house is asleep always encourage me to keep on writing. I hope you enjoyed the book. Thank you to the readers of my *Cumbria* magazine column.

Special thanks to Maria Benjamin and John Atkinson, firstly for allowing my family and I to stay at Parkamoor and experience what it was like living and sleeping up there and secondly to John, Maria and the National Trust for allowing me to use the name Parkamoor to identify the farm. If you would like to stay at Low Parkamoor, the cottage in the clouds, have a look at www.dodgsonwood.co.uk and you can have your own adventure there. I thoroughly recommend it.

Thanks also to Maria for providing some of the photographs of the house, and to Christopher Marshall one of their guests at Parkamoor, for the cover photograph. I offered to send the story to John and Maria before it was published, but they insisted that they should read it at the same time as everyone else. I feel that I should point out again that the farmer in the story is not John, he is a figment of my imagination.

Thanks to the Mason family for allowing me to publish one of the short stories. I felt that it was important that not all the stories were about adults, and not only set on farms, if I was trying to illustrate life in a Lakeland valley. The children living in the valleys have a wealth of knowledge simply through 'lived knowing' and I felt this should be acknowledged.

The phrase 'lived knowing' brings me to Dr Sarah May of the University of London. Sarah worked alongside me for two years while she was researching the concepts of the cultural heritage of shepherding and the idea of rewilding,

as part of the World Heritage Bid for the Lake District. Sarah wanted to see beyond the beautiful scenery and find out what was really important to those living and working in the valleys. I hope that this book has helped other people to understand this too, as this was my aim.

It is very easy to drive through the valleys of the Lake District and see the beautiful scenery and the quaint old farmhouses. Perhaps now the reader can appreciate a little bit more of what goes on behind the farmhouse gates, that there is probably a woman inside despairing because one of her sheep has died and she has no hot water to run a bath!

Finally, thank you to you the reader. I hope you enjoyed the book and found it entertaining. My story will continue in a different valley in Westmorland, as we move 'lock, stock and barrel' and experience the most extreme weather conditions of our lives. See you there.

Andrea Meanwell, July 2018

Thank you to Amy Bateman for the author photograph opposite.

Read More...

books by Andrea Meanwell

A Native Breed, Starting a Lake District Hill Farm,
978-1-910237-24-3
(now out of print but we are taking orders for a new edition)

In My Boots, A Year on a Lake District Farm,
978-1-910237-24-3

www.hayloft.eu